Shedding Light on the Battlefield

Tactical Applications of Photonic Technology

Joseph N. Mait, Michael M. Haney,

Keith W. Goossen, Marc P. Christensen

Portions of this work were supported by a grant
from the National Defense University.

November 2004

Joseph N. Mait is a senior researcher at the US Army Research Laboratory. His research interests are in sensors and the application of optics and photonics to sensing and sensor signal processing. During the academic years 2001-2004 Dr. Mait was assigned to the Center for Technology and National Security Policy at the National Defense University where he researched topics related to Army transformation and technology.

Michael W. Haney is a Professor of Electrical and Computer Engineering at the University of Delaware. His research is in the application of photonics to new computing, switching, signal processing, and image processing architectures. He was previously the Director of Photonics Programs at BDM International, Inc., as well as an Associate Professor of Electrical and Computer Engineering at George Mason University and co-founder of Applied Photonics.

Keith W. Goossen has 17 years professional experience in the fields of optoelectronics and optical fiber communication. While at Bell Labs he invented and demonstrated several key components necessary for high-speed optical communications. In 2000 he co-founded Aralight, Inc. to commercialize this technology and guided engineering development to a full product demonstration two years later. In 2002 he joined the University of Delaware as an Associate Professor of Electrical and Computer Engineering to reengage in the pursuit of technology research.

Marc P. Christensen is an Assistant Professor of Electrical Engineering at Southern Methodist University, where he researches information efficient sensor architectures and processing architectures that use photonic technology. His expertise is in hardware demonstrations of photonic technology gained through his industrial experience as technical leader in Sensors and Photonics at BDM International, Inc. and co-founder of Applied Photonics, Inc. in 1997. He joined Southern Methodist University in 2002.

The authors would like to thank the Center for Technology and National Security for its support of this project.

Defense & Technology Papers *are published by the National Defense University Center for Technology and National Security Policy, Fort Lesley J. McNair, Washington, DC. CTNSP publications are available online at http://www.ndu.edu/ctnsp/publications.html.*

Contents

Executive Summary

The availability of tactical sensor technology places great demands on bandwidth for the battlefield. To relieve some of the bandwidth congestion, we consider using light—or more properly, photons as opposed to electrons— to transmit information. The key advantage of photonics over electronics is the ability of light beams to cross, which allows photonic systems to be more volume efficient than electronic systems. Photonics is a key enabling technology for the military's global information grid, but it has many other applications.

We highlight the advantages of photonics in three applications on a single sensor platform: off-platform communications, thin optical sensors, and on-platform communications. Direct laser communications can provide high-speed, secure communications with an unmanned aerial vehicle using less volume than standard radio frequency technologies. This is due primarily to the smaller wavelength. Further, compact sensors for imaging are possible because photonic and electronic devices can be fabricated using the same conventional fabrication techniques.

For economic reasons, now is the time for the government to take full advantage of photonics. The downturn in the communications sector of the economy at the turn of the century means a large commercial infrastructure is currently underutilized. Capacity, technology, and intellectual property can be leveraged at relatively low cost by the government.

To facilitate the government's leveraging commercial technology, we recommend that the Director of Defense Research and Engineering re-establish the Department of Defense's Photonics conference, which is presently defunct. To facilitate the transition of photonic technology, program managers and others from the acquisition community need to be aware of the technology presently available in the private sector. The advantages of photonics extend beyond just the three applications we highlight here. Photonics is a broad based technology whose impact on sensing, signal processing, and communications may be as great as electronics' impact.

Funding agencies and service laboratory staff are already aware of the advances in photonic technology. What is necessary now is to increase the exposure that photonics has in development centers, system centers, and in the acquisition community. To provide the necessary coverage, the authors propose resurrecting the DOD Photonics series of conferences that existed in the 1990s.

Introduction

Military applications of optical technology have a long history. For over 300 years, telescopes and binoculars have extended the range of a commander's vision. Periscopes allow submariners to view surface activities while submerged and, during World War II, the Norden bombsight provided American bombardiers accuracy that, although crude by today's standards, was unparalleled for its time.

Since the invention of the laser in 1960 and the light emitting diode in 1965, advances in electronics have spilled over into optics and brought opto-electronics to the battlefield. Shortly after its invention, the laser was used to guide munitions in Vietnam. Night vision technology also made its first battlefield appearance in Vietnam. More sophisticated infrared imaging gave coalition forces in Operation *Desert Storm* a critical advantage in night operations. Advances in optics have enhanced air operations with cockpit head-up displays based on the principles of holography.

Communication systems also continue to benefit from advances in optics. In fact, free space laser communication is integral to inter-satellite communication for the Department of Defense's global information grid (GIG). The ground-based portion of the GIG is heavily dependent upon optical fiber to provide up to 10×10^9 bits, or 10 gigabits (Gb) per second data rates.

Miniature lasers and the opto-electronic detector arrays that have replaced film in imaging systems are two notable examples of technologies referred to as photonics. The term photonics derives from the photon, the elementary particle of light. In electronic systems, the electron carries information. In photonic systems, it is the photon. The term photonics is also used to distinguish between systems that use conventional optical elements to form images and those that use light to communicate, compute, and store information. In short, photonics is to optical as electronics is to electrical.

One of the first applications of photonics to communications was the photophone,[1] demonstrated by Alexander Graham Bell in 1880, which used light beams to transmit information wirelessly. Bell believed the invention of the photophone was more significant than that of the telephone, but it took almost a century for light to be used in communication; the first widespread deployment of optical fiber began in the 1970s. The recent downturn in the telecommunications industry was fueled in part by unmet expectations in the growth of optical fiber communications. However, the downturn was due to poor market predictions, not poor technology. The dependence of the GIG on optical fiber indicates strong support for photonics as an enabling technology for transformational communications.

Although the application of optics to military communications is as ancient as warfare itself, the application of photonics is relatively more recent. Nevertheless, as indicated above, it has already had significant impact. Unlike optical technology, photonics cannot function independently of electronics, even though it is the heart of many systems. For example, systems to counter heat-seeking missiles use semiconductor lasers operating in the infrared spectrum and electronics to control the pointing and modulation of the lasers. Similarly, lasers are the core technology in some directed-energy weapons and in guided munitions.

---

1. Bell's photophone used sunlight and vibrating mirrors to transmit the human voice. http://inventors.about.com/library/inventors/bltelephone3.htm, accessed April 19, 2004.

Here we address the application of photonics to sensing and information processing for intelligence gathering, surveillance, and reconnaissance (ISR), in which the efficient generation and delivery of optically-encoded information is exploited. The outer shell of future military networks will be populated by sensors, and the Department of Defense (DOD) is pushing to provide sensor capabilities to tactical commanders. However, in tactical operations, bandwidths are reduced[2] and operational urgency prevents data from being transmitted to ground stations for subsequent processing. To take full advantage of new capabilities, sensors must be able to collect data and rapidly extract from it and transmit actionable information. This can be accomplished if information is generated as close to the sensor platform as possible. However, tactical platforms, e.g., mini-unmanned aerial vehicles, place a premium on the size, weight, and power requirements of a sensor package.[3] Other potential platforms include unattended ground sensors, unmanned ground vehicles, and even dismounted soldiers.

It is in such applications, where the complexity of processing is high and the physical constraints on the system are limiting, that photonics offers the greatest advantage over electronics. The advantage lies in the fact that, whereas two electrons in close proximity affect one another, two photons do not. That is, wires cannot cross, but optical beams can. Anyone who has played flashlight tag has exploited the latter and anyone who has not paid careful attention to the former typically finds him- or herself groping in the dark for a breaker switch.

Because optical beams can share space, the volume needed to transmit data is reduced. This physical difference allows photonic circuits to be more space efficient than electronics. For tactical networking the efficient transmission of data through a volume is critical. Below we consider several applications in which photonics offer a distinct advantage over electronics in meeting this objective.

2. The Army's Bandwidth Bottleneck, (Congressional Budget Office: Washington, DC, August 2003). ftp://ftp.cbo.gov/45xx/doc4500/08-28-Report.pdf, accessed April 13, 2004.
3. T. Coffey and J. Montgomery, "The Emergence of Mini UAVs for Military Applications," Defense Horizons 22 (National Defense University Press, December 2002).

Operational Environment

Before addressing the specifics of photonics, we present a nominal processing application. Consider an unmanned aerial vehicle (UAV) used to collect hyperspectral imagery[4] while operating at an altitude between 3,000 and 10,000 ft. Collecting a hyperspectral image is analogous to photographing a scene using a sequence of red, green, and blue filters in front of the lens. Using the information provided by a hyperspectral imager, an analyst may be able to distinguish between a metallic platform, the trees under which it is hiding, and the wooden decoy placed next to it.

We now consider the implications for data collection and processing using a hyperspectral imager with 20 spectral bands. We assume the data is being collected to locate a target that is 1 foot in diameter, the approximate width of a human.

From the geometry represented in figure 1, if an imaging system has a 5-degree total field of view, in two-dimensions there exist 262×262 1-sq. ft. spots within the system's field of view. To insure accurate detection and recognition, we want this square foot area to fall onto more than one detector pixel. A typical value is 2.5×2.5 pixels. Thus, a single image capable of displaying accurately a one square foot target from 3000 ft. consists of 427716 [$= (2.5 \times 262) \times (2.5 \times 262)$] pixels.

The amount of light collected at each pixel is represented by a gray level that is converted into bits. For example, it is possible to display 256 gray levels using 8 bits ($2^8 = 256$), and slightly more than 4000 gray levels using 12 bits ($2^{12} = 4096$). If we assume the detector has 12 bits, an image in a single wavelength band consists of 5.1×10^6 bits, or 5.1 megabits (Mb). Thus, a single hyperspectral image with 20 spectral bands contains over 100 Mb. Finally, generating a hyperspectral movie, that is, generating a new hyperspectral image 30 times in one second, requires a data rate of 3×10^9 bits per second, or 3 gigabits per second (Gbps), which is 50,000 times faster than the data rate of a 56K modem.

This example is meant to be representative and not definitive. The exact data rate is dependent upon the operation of the UAV, target size, number of spectral bands, detector dynamic range, and update rates. However, these parameters have little effect on the order of magnitude of the data rate. In most proposed applications the data rate rises very quickly to gigabits per second and can easily approach terabits (10^{12} bits) per second (Tbps). The ultra-high information density and data rates envisioned for future sensors systems, in combination with the desire to decrease the size of sensor platforms while increasing their effectiveness, means that data rate bottlenecks will be problematic at all levels of the data collection, processing, and distribution chain. In the following, we consider photonics as an effective means to sustain these anticipated data rates onto and off the platform, as well as within the platform.

---

[4] A hyperspectral image is a collection of conventional images taken over a broad range of wavelengths but using narrow spectral filters.

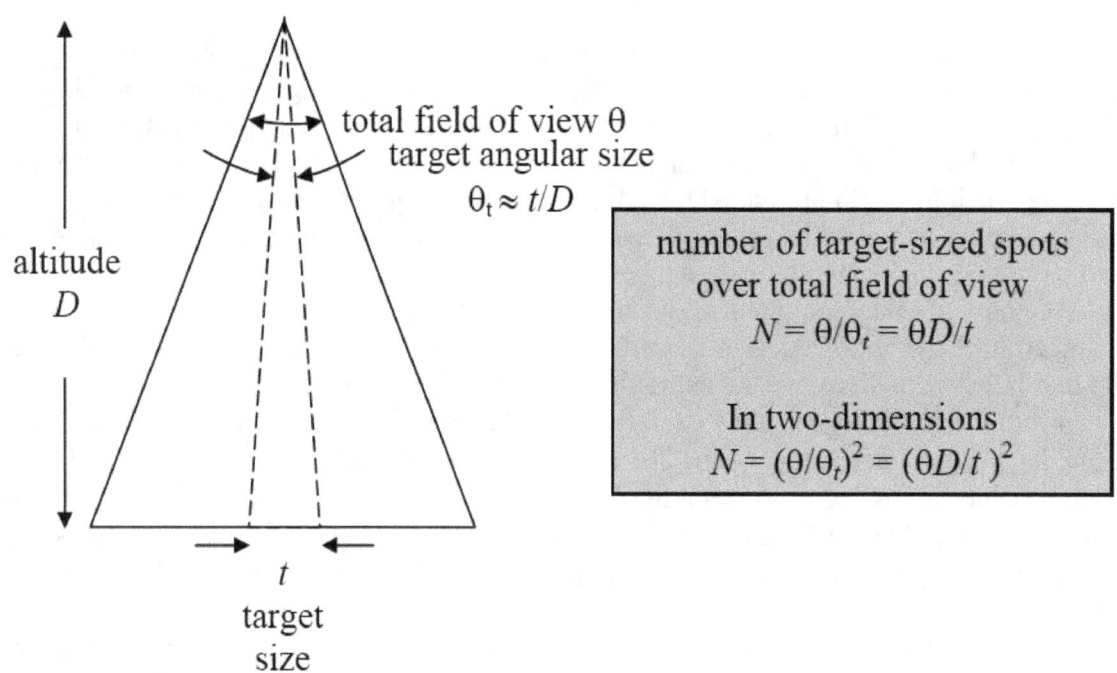

Figure 1. Geometry for imaging from a UAV.

Photonic Applications

We now consider three distinct applications related to the UAV example presented above: communication off the platform, imaging onto the platform, and processing on the platform.

Off-platform Communication

Our UAV example highlights how easy it is for a networked battlefield to generate high data rates. The target capacity for the GIG, the military's planned backbone for global networking, is 10 Gbps, and that is for national assets. Our UAV is a tactical asset. We ignore this constraint for now and consider how to retrieve data from the platform in an efficient and secure manner.

It is important that any transmission to or from the UAV be difficult to detect and intercept. This can be achieved by encoding data so that transmitted signals are essentially similar to noise. Prior to any processing, detected signals should appear as little more than static. However, this encoding requires a portion of the bandwidth available to the platform. The maximum data transmission rate that radio frequency (RF) technologies can achieve in the atmosphere is about 100 Mbps without any secure encoding. With encoding, the data rate is reduced by at least an order of magnitude to 10 Mbps but, depending upon the antenna system, can be reduced to 100 Kbps. Further, it is necessary to assign portions of the transmission frequency spectrum to insure that signals do not interfere. However, this assumes signals are broadcast omnidirectionally, that is, equally in all directions. Enhanced security is possible if signals are transmitted using directional antennas.

Photonic technology can achieve these same capabilities at reduced weight and increased data transmission rates. The higher frequency of light as compared to radio frequencies and the low-loss propagation of light through the atmosphere provide a substantial increase in the bandwidth available for communication. (The smaller wavelength also reduces the size of sources and detectors.) Although the system does not operate well in rain or fog, the hyperspectral sensor described above is similarly affected. Thus, weather conditions that permit the use of a hyperspectral sensor also permit the use of optical communications.

If the UAV is operated either autonomously or semi-autonomously, photonic communication offers a simple means for transmitting data off the platform. In semi-autonomous operation, in which the UAV is tele-operated, RF technology can be used to up-link control signals and photonic technology can be used for the data downlink. The downlink can be constructed simply using a ground-based laser, a corner-cube mirror on the UAV (a corner-cube consists of three planar mirrors arranged to form the interior corner of a cube), and a modulator on the UAV that can impart electronic data on top of an optical beam. When the beam from the ground-based laser hits the corner-cube, it is directed back onto itself. Thus, no matter at what angle one looks into a corner cube, the viewer will always be looking directly into his or her face. So long as four corner-cubes are used on the UAV (one for each quadrant of the platform, fore and aft, port and starboard), only the laser needs to be scanned to insure that it follows the track of the UAV. The modulator encodes the sensor data on the optical beam by controlling the beam's intensity.[5]

Again, this type of free space optical communication system provides directional communications with minimal requirements on power. It is ideal for a downlink, because the UAV does not have to generate an optical signal, but merely reflect it. In energy-limited communication, it is assumed that the ground-based receiver has plenty of energy, while the UAV does not. The size and weight of the mirrors and modulators can be made relatively small. In fact, using multiple quantum well (MQW) technology,[6] or micro-electro-mechanical systems (MEMS),[7] a payload designer can keep the weight down to only a few grams. The payload designer also needs to consider the energy required to drive the mirrors and modulators, and to format or compress signals for transmission.

In some situations (for example, an unattended ground sensor) the sensor platform is not transmitting continuously. In these cases, it is possible to modify the link described above so that the sensor responds only when a base station queries it optically using, for example, a binocular-like device equipped with a laser pointer.[8] The laser pointer triggers the sensor and the collected data is displayed on the binocular. Note that, under these conditions, rain and fog will affect communication. The concept of operations therefore needs to be considered carefully. Also unlike the UAV-mounted hyperspectral sensor, the ground sensor must detect the presence of an interrogating beam before transmitting data. If we assume the sensor and interrogator are not moving, the system can be designed without corner cubes. Instead, the sensor is equipped with its own array of lasers and an array of adjustable mirrors. The sensor determines the direction of the interrogating beam and uses the array of adjustable mirrors to direct the sensor's laser in that direction. The laser is modulated directly by the sensor data. To keep size, weight, and power minimal, semiconductor lasers must be used, for example, vertical cavity surface emitting lasers (VCSELs), which can be fabricated in arrays.[9,10] In addition, such a system requires a wide range of other photonic technology, including arrays of MEMS mirrors and detectors. It also requires intimate integration of these technologies with electronic circuit technology to perform the computation of the incoming beam direction and resulting actuation of the mirror array.

5. G. C. Gilbreath, W. S. Rabinovich, T. J. Meehan, M. J. Vilcheck, R. Mahon, R. Burris, M. Ferraro, I. Sokolsky, J. A. Vasquez, C. S. Bovais, K. Cochrell, K. C. Goins, R. Barbehenn, D. S. Katzer, K. Ikossi-Anastasiou, and M. J. Montes, "Large-aperture multiple quantum well modulating retroreflector for free-space optical data transfer on unmanned aerial vehicles," Optical Engineering, vol. 40, pp. 1348-1356 (2001).
6. W. S.Rabinovich, R. Mahon, P. Goetz, E. Waluschka, D. S. Katzer, S.Binari and G. C. Gilbreath, "A Cat's Eye Multiple Quantum Well Modulating Retro-reflector," IEEE Photonics Technology Letters, vol. 15, pp. 461-463 (2003).
7. C. Luo and K. W. Goossen, "Optical micromechanical system array for free space retrocommunication," IEEE Photonics Technology Letters, vol. 16, pp. 2045-2047 (2004).
8. Private communication with Prof. Joseph Ford, University of California at San Diego.
9. L. Buckman Windover, J. N. Simon, S. A. Rosenau, K. S. Giboney, G. M. Flower, L.W. Mirkarimi, A. Grot, B. Law, C.-K. Lin, A. Tandon, R. W. Gruhlke, G. Rankin, and D.W. Dolfi, "Parallel-optical interconnects greater than 100 Gb/s," J. Lightwave Tech., vol. 22, pp. 2055-2063 (2004).
10. C. Cook, J. E. Cunningham, A. Hargrove, G. G. Ger, K. W. Goossen, W. Jan, H. H. Kim, R. Krauss, M. Morrissey, M. Perin, A. Persuad, G. Shevchuk, V. Sinyanski, and A. V. Krishnamoorthy, "A 36-channel parallel optical interconnect module based on Optoelectronics-on-VLSI technology," IEEE J. Selected Topics in Quantum Elec., vol. 9, pp. 387-399 (2003).

Thin Attentive Optical Sensors

In the previous section we considered photonic technology for moving data at high transmission rates off a platform. However, it is also possible to use photonic technology to improve the collection of imagery by the platform without relying upon traditional optical solutions.

To understand this, it is necessary to point out that the diameter of the optical elements in an imaging system directly affects the amount of light the system can collect. Thus, to image faint objects one uses a large lens. Lens diameter also affects the system resolving power. That is, the lens diameter determines the smallest object the system can discern as a single object. Large lenses are also used to form high-resolution images. However, it is a physical fact that large lenses have long focal lengths. Consequently, despite advances in the miniaturization of electronics, imaging sensors remain bulky cubes. Further, costs associated with the design, manufacturing, and packaging of these bulky complex systems have made them a relatively scarce resource. To make imaging resources pervasive, imaging sensors must deviate from the typical bulky cube so they can be readily integrated into a variety of military and personnel systems. A flat imaging sensor is an obvious choice.

One only needs to consider the recent revolution in imaging displays to gauge the possibilities for a flat imaging sensor. The state of the art in display technology is flat. Whether it is plasma television sets or liquid crystal display (LCD) computer monitors, flat display technology has become prevalent. Only through the application of new technologies was the bulk display paradigm of the cathode ray tube replaced.

A shift toward flat imaging sensors is possible if one considers that, in the traditional optical imaging paradigm, a lens forms an image onto a solid-state detector array. In radar and radio astronomy, however, incoming radiation is collected directly without a lens and images are formed by applying computer algorithms to the collected data. These two systems represent the extremes in image formation. The purely physical means of the optical system produces large-sized systems. But the purely computational means of radar generates large data sets and large computational loads. The ideal tradeoff lays in between these two, where optics is used to condition the input in such a way that the computation required to form the image is reduced.

So long as the optics is not used to form the final image, it is possible to create a thin imaging sensor using small optical elements, for example, a square centimeter array of lenses whose diameters are smaller than one millimeter. Additional electronic computation is required to form the final image. We note that the imager detects intensity only and uses the multiplicity of images produced by the lenslet array to generate an enhanced resolution image in post-detection. The imager is not based on interferometry.

Flat imaging sensors based on arrays of micro-optical elements have already been demonstrated.[11] It is important to point out that, although a quick Internet search turns up many thin cameras, all use conventional optical engineering techniques to fold the optical system. That is, the distance required for image formation is placed in the width of the camera, not its depth. The imaging system we describe is thin in width and depth, and uses the novel balance between physics and computation to produce an image.

---

11. J. Tanida, T. Kumagai, K Yamada, S. Miyatake, K. Ishida, T. Morimoto, N. Kondou, D. Miyazaki, and Y. Ichioka, "Thin Observation Module by Bound Optics (TOMBO): Concept and Experimental Verification," Appl. Opt., vol. 40, pp. 1806-1813 (2001).

Once the traditional image formation paradigm is broken, it is possible to apply photonic technology beyond micro-lenses to provide additional imaging capability. For example, micro-lens arrays used in combination with steerable micro-mirror arrays, similar to those used in many laptop projectors, can be used to generate images with varying resolution. Imagine initially acquiring an image that has uniform resolution. If we assume computational resources are placed close to the sensor, it is possible to use low-level processing to highlight regions of interest. The location of these regions can be fed back to the imaging system to adjust the micro-mirror arrays so that multiple lenses are used to image the same region. In this way it is possible to increase the resolution in the regions of interest and decrease it elsewhere without changing the orientation of the entire imaging system.

Such a system is an attentive multi-resolution imager. Image resolution is variable across the image and it is capable of adapting continuously to the scene. These capabilities have a definitive impact on the data collected by the platform and its associated bandwidth. Instead of relying solely upon data compression algorithms to reduce bandwidth, information compression is achieved using an adaptive collecting system. This adaptive approach can actually outperform the high-resolution bulk imaging sensor if the entire scene is not of interest–as is typically the case for tactical scenes. Again, the key is balancing the processing between the physical and computational domains, which can be achieved by exploiting adaptive optical elements, electronic processing, and feedback from the processing to control the optics.

If our UAV is tiled with flat imaging sensors, it can survey an entire battlefield instead of relying upon a single optical sensor that must be mechanically zoomed on a single area of interest. Flat imaging sensors mounted to a soldier's helmet can report data not only to the soldier but also to commanders, all without adding strain to the soldier's neck or hindering the soldier's movements. Hallways tiled with attentive flat sensors provide an additional layer of physical security. Since the sensors can have a large field of view, an intruder is unable to hide "behind the camera." Form factor is the single greatest obstacle to prevalent image information, and technologies have emerged that enable a shift in imaging paradigm.

On-platform Communication

In previous sections we considered how photonic technology could be used to enhance data flow onto and off a platform. In this section we show how photonics can enhance the flow of data on the platform itself.

As digital silicon integrated circuit (IC) technology continues its exponential growth in speed and density of devices, raw computational power continues to grow at an astounding rate. The full exploitation of silicon's computational power, however, requires that the links between devices and sub-systems keep pace with increasing bandwidth. As stated, future networks depend on the efficient transfer of data through a volume. This includes the transfer of data within the processors that enable the network.

Consider the data flow required to facilitate rapid, real-time decision-making. To produce the information necessary for decision-making, data must travel between high throughput sensors, such as optical imagers or radars, data storage elements that store target templates or reference imagery databases, and multiple processing elements. The requirement that data be delivered in as short a time as possible places high demands on the communication pathways between these modules. Historically, however, the performance of metal-based interconnect technology has lagged computing technology. Put simply, bus speeds are always slower than processing speeds.

It is important to understand that data in silicon-based processors moves between components that differ in physical scale. The components are, from largest to smallest, the cabinet, module, printed circuit board, chip carrier (the recognizable black ceramic package), and, finally, the chip, which is mounted inside the chip carrier. Although deficiencies exist at all levels in transmitting data efficiently between components of the same scale, e.g., between circuit boards within a computer, the problem is exacerbated for communication between layers. This is due to the mismatch in bandwidth densities at each level. (Bandwidth density is a measure of how much information can flow through a physical boundary, such as the back of an electronics cabinet, measured in square meters (m^2), or the edge of an IC, measured in centimeters (cm). Hence, depending on the interconnection geometry in question, bandwidth density can have units such as Gbps per m^2 or Gbps per cm.)

The electronic-metal interconnection problem stems from practical limits on wire size, spacing (or pitch), and speed. The smallest wires, lowest pitch, and highest speeds are available on the chip, where size and pitch are measured in micrometers (10^{-6} meters) and speeds in excess of 10 Gbps per wire are possible. To communicate with modules in the next packaging level the chip must have some metallic path to, for example, a chip carrier or printed circuit board. This is provided by metal nodes approximately 50 to 100 micrometers in size with 125 to 150 micrometers between nodes. The nodes operate at speeds typically less than 1.0 Gbps. There is, therefore, a fundamental mismatch in communication (i.e., bandwidth density) between the chip and the outside world. This mismatch is repeated at each level of a system packaging as wire size and pitch increase with increased scale.

Consequently, high-performance systems tend to be *communication-limited*, rather than *computation-limited* in their performance. The limitations of metallic interconnects at the cabinet, module, board, and carrier levels are already affecting processor performance.

Although interconnect technology continues to be dominated by metal and electronics, photonic technologies have been encroaching on electronics for over 25 years. This trend can be traced back to the introduction of optical fiber into telecommunications in the late 1970s. Optical fiber has shown that photonics technology can provide revolutionary solutions to bottlenecks in long-distance communications.

Since the 1970s, photonic interconnects have moved continuously into applications with ever-shorter path lengths and are presently making significant inroads in communication between cabinets and between modules inside the cabinet. For example, in the mid 1990s Motorola developed the first fiber interconnect for box-to-box interconnects that relied on micro-laser arrays to drive data through a ribbon of parallel optical fibers. The ribbon of 12 fibers provided nearly 10 Gbps of aggregate capacity. Since then, individual channel rates have risen to 10 Gbps, and aggregate capacity has increased by an order of magnitude to 100 Gbps.

To extend photonic interconnects to even shorter length applications, major research efforts are now underway to use photonics for communicating between chip-carriers and other components on a circuit board. Communications on a circuit board are already experiencing bottlenecks due to limitations of metal-based technology.

Although the impetus from the scientific community to use photonic interconnects at decreasing length scales has been consistent for more than two decades, the technical challenges at each length scale have differed. To transmit data between cabinets and over long distances using optical fiber required discrete optical sources and detectors (or small arrays of approximately 10 elements) that operate at high power, at high modulation rates, and within narrow optical wavelengths. These needs have been met.

In contrast, due to the sheer density of transmitted signals on a circuit board, to apply photonic interconnects to board-level connects requires significant improvements in miniaturization, integration, packaging, reliability, and power efficiency. It is noteworthy that the most successful demonstrations of photonics are those that leverage the scaling advantages of silicon-based technologies in lithography, integration, and packaging.

Preliminary research in packaging arrays of optical sources and receivers with optical waveguides (roughly the equivalent of optical wires in a piece of glass) show promise for board-level interconnects. Waveguide approaches have exhibited higher bandwidth density and efficiency than metal-based competitors. Thus, in 2002 DARPA initiated the chip-to-chip optical interconnects (C2OI) program to develop board-level photonic interconnects between silicon chips using imbedded optical waveguides, instead of metallic traces. The goal of the program is effectively to erase the bottlenecks between chips so that chips mounted on a board can communicate at *on-chip* speeds. If successful, the C2OI concept could provide one to two orders of magnitude improvement in computations for a given system weight and power consumption.

Over the past two decades, photonic interconnect technology has been applied successfully to transmission lengths that vary over six orders of magnitude. In the long-haul regime, wavelength division multiplexing has been successfully applied to provide signal transport, routing, and switching over hundreds of kilometers. At the other end of the length spectrum, the reliable integration of photonic devices in dense packages is enabling chip-to-chip interconnects over tens of centimeters (the goal of the C2OI Program). The obvious next step is to extend the length-scale trend yet another order of magnitude to the inside of a chip, the *intra-chip* domain, where interconnect lengths are about a centimeter.

The motivation for this stems from the limitations of metal interconnects for high-speed data transfer in densely populated silicon ICs. Moore's Law addresses only with the doubling of transistor density on a chip, not the rate at which the transistors can communicate. As the density of transistors increases, the difficulty in designing the interconnection fabric also increases.

To understand the problem better, consider designing the roads for a neighborhood in which a road must exist between many pairs of homes but no part of any road can be shared with any other road. (Recall that wires cannot cross.) If the neighborhood is sparsely populated, this is not particularly difficult. However, the design becomes more complicated as the density of homes increases. In fact, the real estate required for the traffic system (the interconnect fabric) may far exceed the real estate for the homes. The technological advances that drive Moore's Law allow the widths of the roads to become narrower without impeding traffic flow but at the same time, these advances also allow more homes to be put on the same area, which does impede flow.

The severest problem arises from the longest roads. If the roads are too long, the neighborhood requires gas stations to insure that cars can reach their destination. Not only do the stations require additional real estate, but the time required for each fill-up also increases the total time of the trip between two homes, which reduces average speeds.

This simple transportation analogy explains why the interconnections between chip components are a critical bottleneck. The complexity of the problem can be reduced if paths are allowed to cross (which is possible with optical signals). This reduces the real estate required, reduces the length of the longest path, and, consequently, increases the speeds at which traffic moves.

The problem facing silicon circuit performance is represented in figure 2, which shows the projected *required* on-chip global bandwidth density (measured in Tbps per square centimeter) as a function of feature size in a silicon IC.[12] These requirement projections are based on an extension of the current microprocessor chip design paradigm. This is compared to the projected *achievable* densities at each technology feature size. When feature sizes approach 65 nanometers (10^{-9} meters), which is forecast to occur about 2007, the bandwidth required to provide communication on the chip is greater than what can actually be achieved. For minimum feature sizes less than this, unconventional methods are required to meet interconnect needs on the chip. If one considers the power required to provide communication, the problem is actually worse than figure 2 suggests. The fundamental nature of this problem, and the impact it will have on all future high-performance silicon chips, provides one of the best opportunities for a photonic solution in which global electronic signals are converted into optical signals and routed over an optical interconnection fabric.

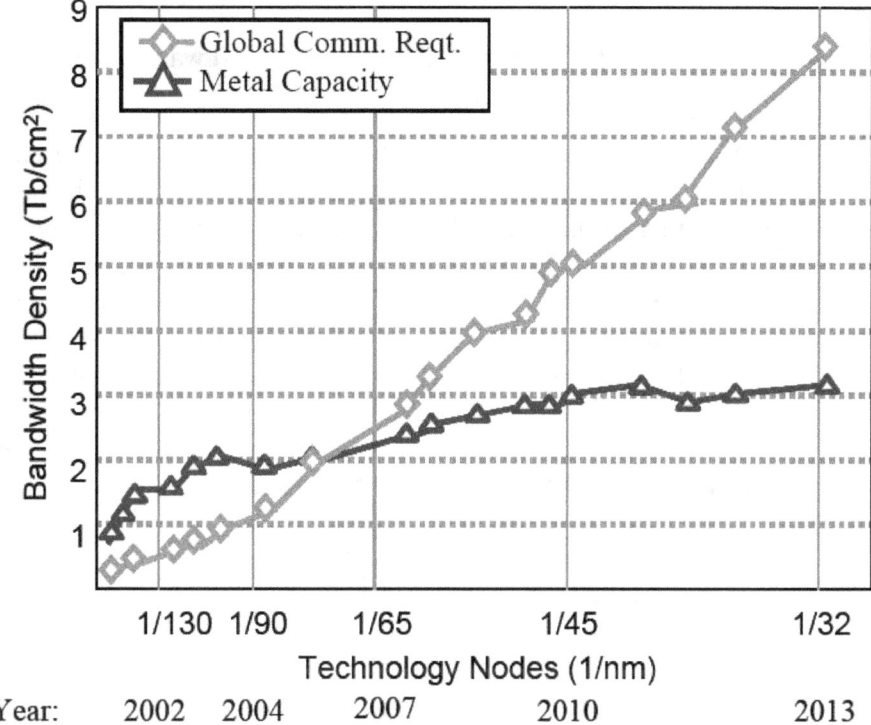

Figure 2. Global wire bandwidth requirements and capacities for intra-chip electrical interconnect fabrics as a function of projected feature size.

Figure 3 depicts two basic technologies for creating chip-scale photonic interconnects: guided wave and free-space. In both cases, electronic signals in the silicon IC are transformed into optical signals by micro-emitters, whose beams are channeled vertically via micro-optical elements (depicted as lenses in the figure) into the interconnection fabric. In guided wave approaches (figure 3a), global metallic wires are replaced by optical waveguides. These planar structures mimic the metallic global wiring layers that they replace—except actual path crossing points can be achieved

12. M. Haney, M. Iqbal, M. McFadden, and U. Hameed, "Intrachip Optical Interconnects: Challenges and Possible Solutions," 2004 ICO International Conference: Optics and Photonics in Technology Frontier, Tokyo, July 2004.

because optical signals can pass through each other without interacting (unlike electronic signals on metal wires). Fewer physical global wiring layers would therefore be needed. In a free-space architecture, guide structures are not required (figure 3b). Instead, arrays of customized micro-prisms and a mirror are used to direct optical beams between the source and destination points on the chip.[13]

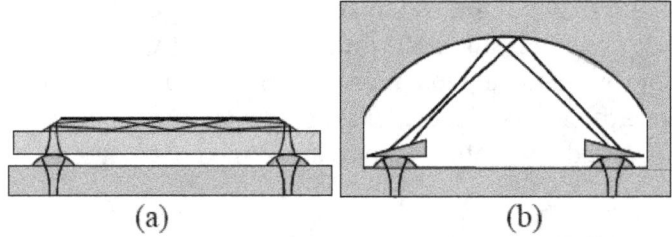

(a)   (b)

Figure 3. Two notional optical interconnect fabrics: (a) guided wave, (b) free-space.

Preliminary examinations of guided-wave and free-space arbitrarily configurable configurations suggest that optical interconnection fabrics may be a solution to the limits on on-chip communication. Figure 4 summarizes these results. The projected bandwidth capacities for the two global optical interconnect technologies are overlaid on top of the requirements for future silicon generations from figure 2. These projections are based on an analysis of the physical constraints on the optical fabrics and an assumption that photonic transceiver arrays will be available that match the silicon line speeds. The data suggest that both guided wave and free-space fabrics are candidates for overcoming the limitations on global on-chip interconnection. Thus, the use of optical interconnection fabrics may delay the performance crossover point between requirement and capability for another decade beyond what metal fabrics can achieve.

13. M. W. Haney, M. J. McFadden, and M. Iqbal, "An Application Specific Interconnect Fabric (ASIF) for free-space global optical intra-chip interconnects," Technical Digest of the Optics in Computing Topical Meeting, June 2003, pp. 105-107.

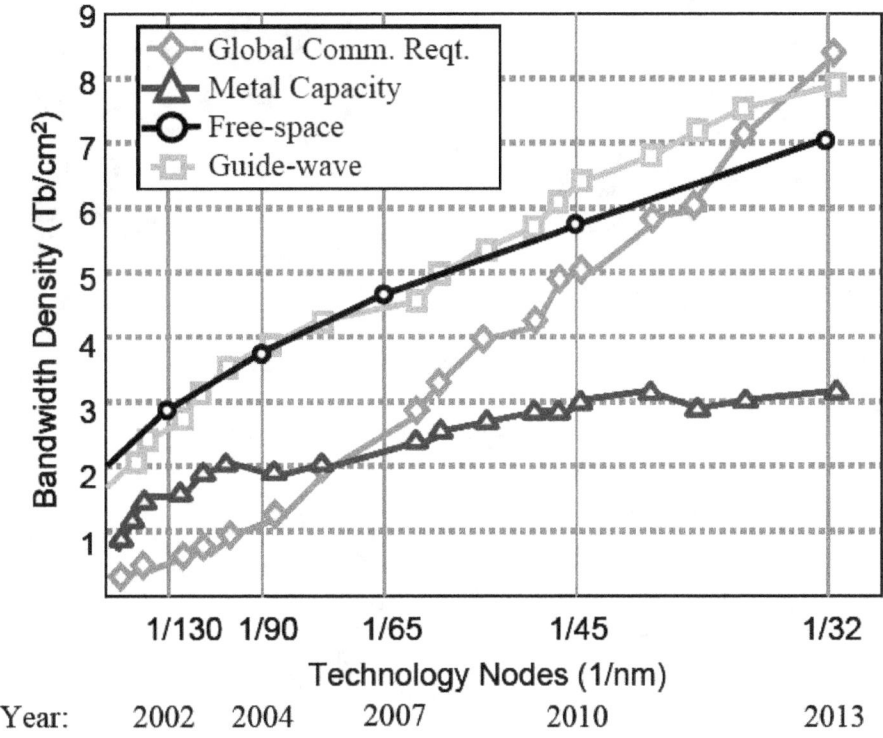

Figure 4. Projected global intra-chip interconnection capacities for photonic fabrics, showing that optical interconnects may provide the solution to overcoming the imminent limitations of metal interconnects as the silicon feature size (technology node) shrinks.

Technology Development Required

With regard to integration, economy of packaging, and usage, photonics technology lags electronics. The reasons for this are based in both physics and engineering. Consider that the integration of several transistors on a single chip is achieved by repeating many times over the chip area the pattern of different materials that make a transistor. The patterns are produced using photolithography, that is, transistors are created through successive exposures of photosensitive materials deposited on top of silicon. The same procedure, in fact, can also easily integrate several photonic devices, for example, a laser and a detector, on a single chip. Thus, photonic device fabrication relies upon the same processes as electronic.

The difference between electronics and photonics is how devices are connected; electrically in the electronic chip and optically in the photonic chip. As simple as this may seem, the implications in terms of fabrication and packaging are dramatic. An electronic circuit requires only an electrical conductor, a metallic line, between the transistors. Recall, though, that this requires building a "bridge" at some layer above the one on which the transistors are located. The transistors exist on the lowest level within the circuit, on the "ground floor." Thus, in addition to building the bridge itself, a vertical path from the ground floor to the bridge is required. These vertical paths are the on- and off-ramps to the bridge, so to speak.

Electrical ramps are made by etching a hole (referred to as a via) through the insulating layer that separates the wire bridge from the transistors and filling it with a metal plug. For data rates less than several gigabytes per second electrical signals easily follow the metal path formed by the via plug. However, given the size of the vias, electrical signals in excess of 10 Gbps have difficulty making the 90-degree turns necessary for data flow. (Electromagnetic fields experience a power loss at corners.)

In a photonic circuit, it is straightforward to make the bridge out of an optical waveguide, instead of a wire, but creating an optical path from the ground floor to the bridge is complicated: the direction of the light beam must be changed physically by 90 degrees. Doing so may seem trivial using, for example, a tilted mirror. However, fabricating the mirror within alignment tolerances that insure the optical beam is properly directed is presently a daunting task.

Further, unlike electronic circuits, in which the movement of electrons is perturbed only marginally as they move from a transistor to a wire, the interfaces in a photonic circuit have a more dramatic effect on the photons. Consider that photons created from inside a laser are launched into either free space or a waveguide and, after careful guiding, are absorbed into a detector. The laser, propagation medium, and the detector represent three different material systems. This has considerable impact on the fabrication and integration of photonic circuits. As such, fabrication and integration technologies need to mature before photonic circuits can be applied broadly.

For comparison, to package an electronic circuit, one mounts a chip on a board and attaches wires between the chip and the board. This procedure can be automated so that several bond wires are attached per second. No analogous packaging procedure exists for photonic circuits.

Procedures do exist for packaging photonic circuits in free space using, for example, holograms. However, coupling a waveguide on a chip to an optical fiber remains a tedious and precise procedure, and typically a manual one. Presently, almost all optical packaging is done by hand alignment.

Thus, in terms of fabrication technology, the advancement of photonic circuits requires the development of low loss vias and packaging technology that takes advantage of the ease with which electrical wires can connect and the ability of optical wires to cross.

Finally, the development of low power, high-density photonic transceivers that can exploit the potential pay-off of chip-scale optical interconnection fabrics remains a critical, unsolved problem. Conventional packaging and cooling techniques limit total power consumption to less than about 100 watts (W) per chip. Projections for the global wiring requirements in future IC technology suggest that several thousand global wires will be needed for microprocessors. Limiting the power budget to, for example, 20 W for the chip's global interconnect power consumption for a 2×2 square centimeter chip populated by 1000 photonic devices per square centimeter, requires devices that dissipate only 5×10^{-3} W per channel. This is currently not possible.

Further, although no fundamental barriers exist to achieving high speed, dense arrays of optical transceivers (a combination of photonic source, detector, and associated drive electronics) with this level of power consumption, considerable challenges nonetheless remain. For example, it is not sufficient just to produce arrays of transceivers but to produce reliable arrays. It is also necessary to package arrays in a manner that complements the underlying silicon IC technology. The density of arrays that can be manufactured is limited by the power consumption and reliability of micro-laser technology. Present technology limits array density to levels well below what is required for high density interconnects. Thus, although on-chip photonic communication offers considerable advantage over electronic communication, its potential has not yet been fulfilled and new photonic transceiver technology is required.

Electronic Alternatives

It is our contention that the significant signal loss and degradation that occur in electrical interconnects as signal speeds exceed 10 Gbps justifies the exploration of optical interconnections. It is also our position that future bandwidth demands can be addressed by increasing carrier frequencies into the optical domain. However, our position is not accepted universally. For example, to account for electronic transmission losses and degradation, one can, alternatively, distort data prior to transmission and process it after detection. Further, the bandwidth bottleneck can be alleviated by compressing data. That is, rather than increase the size of the data pipe, one can reduce the amount of data flow. It is clear then that the primary competitor to photonics is not an alternative physical technology, but the application of advanced signal processing techniques.

However, there are limits to signal processing alternatives. For example, compression can only go so far. One should recall the videophones from the 1990s and their inability to produce satisfactory imagery. These phones were designed to operate on 56 Kbps phone lines using compression.

The limitations of pre- and post-transmission signal processing are more subtle to determine because the physics of electronics make them difficult to quantify. One would like to compare either the signal-to-noise ratio or bit error rate of a photonic interconnect to that of a signal processing-enhanced electronic interconnect. This would highlight any advantage of one system over the other. However, the electro-magnetic interaction between electrons complicates the comparison. Indeed, an accurate analysis of an electronic system must consider the influence of every transmission path on every other path because the quality of data transmission in one path is affected by the state of its neighboring paths.

Since photons do not interact, one can consider each photonic path independent of the others. The advantage of photonics over electronics is apparent once again.

If one steps back for a moment to consider the implications of using signal processing to enhance electronic transmission, a circular argument begins to emerge at some point. Consider that, to insure robust high-speed electronic transmission of information, one first requires high-speed signal processing. However, at some point, no signal processing can be performed because the data rate will exceed the physical capacity of a signal-processing chip to accept or transmit signals. Seen from this perspective, the data rate is limited by the physics of a particular link. For example, the most typical link is a bond wire that connects the chip's processor to the outside world. In more advanced modules, the limiting link is the interface between two chips placed one on top of the other. This link, known as a flip-chip bond, has been evaluated extensively[14] and analysis indicates it fails due to signal degradation and power loss when data rates are in excess of 10 Gbps. It is at these data rates and beyond that we expect photonics to make significant inroads into interconnect technology.

Leveraging Commercial Technology

It would be misleading not to mention the research and development efforts in photonics at the service and national laboratories. These laboratories have made significant contributions to the development of miniature lasers, modulators, and electro-optical detectors, and to the development of unique optical components for splitting and shaping optical beams. However, much of this effort is concerned primarily with sensing, including imaging, with some work on signal processing. As indicated here though, and in the 1998 report of the National Research Council's Committee on Optical Science and Engineering (COSE),[15] optics and photonics have potential in countless other applications.

The intent of the COSE report was to broaden the technology base by garnering government support for photonics industries. However, it did not have the same impact in the United States as it did overseas, for example, in Germany, Scotland, France, Korea, Singapore, and Canada. The timing of the report was indeed prescient; it was published just before the technology boom. Instead of government support, billions of dollars of venture capital were invested in photonics.

The government's slowness to respond in the late 1990s essentially allowed the commercial market to finance development of a photonics infrastructure. This was fortuitous if not deliberate. The communications industry's economic implosion slowed the insertion rate of photonic technology into the commercial arena, thus the government can benefit from the large private investment and purchase underutilized capacity and technology at a relatively low price. The GIG, as well as the military applications discussed here, will likely be among the first beneficiaries.

However, the opportunity to leverage this private investment should be seized without haste. The largest portion of the investment went to start-up companies that did not publish,

---

14. A. V. Krishnamoorthy and K. W. Goossen, "Optoelectronic-VLSI: Photonics integrated with VLSI circuits," IEEE J. Selected Topics Quantum Elec., vol. 4, pp. 899-912 (1998).
14. Harnessing Light: Optical Science and Engineering for the 21st Century (National Academy Press, Washington DC, 1998).

patent, or otherwise share knowledge derived from their enterprises. The imperative was "time to market" and not leaving tracks. Thus, a vast source of untapped intellectual property lies dormant and is at risk of disappearing as the individuals who retain it are retrained for jobs in other sectors or retire. There exists only a relatively short window of opportunity to leverage the vast investment made. Once closed, much of the innovation derived from practical "know how" could be lost.

A few non-government programs do exist for transitioning photonic technology. For example, Infotonics (http://www.infotonics.org/), a center of excellence based in upstate New York, promotes commercialization of photonics through prototyping and developing pilot production lines for products. Formed in 2001, Infotonics is an industry-led not-for-profit consortium of industries and universities. It is housed in a 123,000-square-foot former Xerox manufacturing plant, which, once refurbished, will be capable of fabricating, packaging, and testing photonic devices. Infotonics works with both state and federal government agencies to develop promising photonic technologies.

The Optoelectronics Industry Development Association (OIDA) also supports the development of photonic technology through its Photonics Technology Access Program (PTAP) (http://www.oida.org/PTAP/). However, PTAP, unlike Infotonics, focuses on providing pre-commercial photonic technology to the academic community. PTAP compensates industry for supplied devices and allocates the devices to researchers based on competitively evaluated proposals. PTAP is sponsored by the National Science Foundation and DARPA.

Recommendation

Photonics is only one of many technologies that can give the U.S. military a tactical and strategic edge. However, the push to deploy the GIG when the commercial infrastructure in photonics is running below capacity provides DOD with an opportunity to capitalize on its investment. Indeed, in 2002, members of two prominent scientific professional societies in optics and photonics, the Optical Society of America (OSA) and SPIE-the International Society for Optical Engineering, sought to propose a national initiative on photonics through the White House Office of Science and Technology Policy. The intent was to model the initiative after President Clinton's initiative on nanotechnology. However, a proposal never went forward. One of the difficulties in garnering support for such an initiative is that the applications of photonic technologies are broad, and much of new science in photonics is actually covered under the nanotechnology rubric.

Although a presidential initiative in optics and photonics was not proposed, federal interest in the area continues, as evidenced by the October 2004 meeting of the Congressional Research and Development Caucus.[16] The caucus, co-chaired by Reps. Rush Holt (D-New Jersey) and Judy Biggert (R-Illinois), provides Members of Congress and their staff a forum on issues in basic and applied research investment. The focus of the October meeting was "Harnessing Light for America" and considered the applications of photonics to medical imaging.

If we take a cue from the photonic initiative efforts, the key is to highlight potential applications for the right target audience. The UAV scenario highlighted here is only one in which photonics can provide increased capability. Sensing, signal processing, and communication have broad application for the military, and all can benefit from photonic technology. Funding agencies and service laboratory staff are already aware of the advances in photonic technology. What is necessary now is to increase the exposure that photonics has in development centers, system centers, and in the acquisition community.

To provide the coverage we seek, we propose resurrecting the DOD Photonics series of conferences that existed in the 1990s. The DOD Tri-Service Photonics Coordinating Committee sponsored the conferences with cooperation from the Department of Energy and NASA. The last conference occurred in 1996,[17] just before the technology boom. Topics covered then included communications, interconnects, and the sensing of chemical and biological agents. These topics are just as relevant today, if not more so, given the increased emphasis on homeland security subsequent to the September 11 terrorist attacks. Sensing and surveillance are important to both the military and homeland security.

We propose that the office of the Director, Defense Research and Engineering (DDR&E) host the new series of conferences in cooperation with the Department of Homeland Security, the Department of Energy, and NASA. However, it is most important to target the right audience. OSA and SPIE already sponsor conferences for funding agencies and researchers. The focus here needs to be exposing and educating a broader community on the available capacity in photonics

---

16. http://www.bioworld.com/servlet/com.accumedia.web.Dispatcher?next=bioWorldHeadlines_article &forceid=33793 (accessed October 14, 2004).
17. Proceedings of the 5th Biennial Department of Defense Photonics Conference, McLean VA, March 26-28, 1996.

and recent advances. The audience should come from program executive offices and the Office of the Secretary of Defense, including Assistant Secretary of Defense for Networks and Information Integration, Under Secretary of Defense for Acquisition, Technology, and Logistics, and Deputy Under Secretary of Defense for Advanced Systems and Concepts. The Armed Forces Communications and Electronics Association is an appropriate host for the conference series.

We believe now is the time to revive this venue to increase the dialog between technology providers and system developers. Familiarity with and participation in photonic transition programs such as Infotonics and PTAP will also help close this gap between these two groups.

Summary

The key advantage of photonics over electronics is the ability of light beams to cross. The most revolutionary advances stemming from photonics are those concerned with improving communication within electronic processors themselves. If the space utilization advantages of photonics over electronics are exploited fully, it may be possible to delay by five to ten years the inevitable "end" to Moore's Law, when linewidths in electronics are so small that quantum effects dominate electrical behavior.[18]

Photonic wavelengths also provide a compact means to communicate with sensor platforms. New paradigms for designing imaging systems that exploit the combined processing power of optics and electronics should also lead to new, more compact and more functional systems. The technologies for these applications are already under development and could be transitioned to the warfighter within five years.

In 2003, the Congressional Budget Office noted the bottleneck in battlefield bandwidth as advanced communication and ISR capabilities are pushed down to the tactical level of warfighting. To mitigate the bottleneck, the report suggests greater distribution of advanced radio systems than planned and a reduction in bandwidth intensive activities, such as video teleconferencing and maintaining UAV videostreams on a network. Although we are not suggesting the application of photonic technologies will eliminate the need for these measures, we are suggesting that photonics may offer unconventional solutions in communication, image collection, and processing that can alleviate some of the constrictions on bandwidth.

Unfortunately, whereas the great advantage of photonic circuits is that optical beams may cross, this lack of interaction also makes it difficult to connect photonic devices. Investments in photonic packaging technology, including fabrication and integration, are required before the advantages of photonics can be fully realized. However, conditions in the marketplace are such that a small government investment at this time could reap large returns by leveraging the infrastructure and intellectual property created through private investment.

At the beginning of the telecommunications revolution 40 years ago, Bell's claim that the significance of the photophone exceeded that of the telephone no doubt seemed quaint. Developments in the intervening years, however, have strengthened his claim, which was underscored by the NRC COSE report. Yet the advancements should not be as surprising as they first appear. Fabrication technology is already capable of producing features that are several hundreds of nanometers in size, or approximately the wavelength of visible light. As the technology moves towards nanometer-sized features, our ability to control lightwave radiation increases.

The twentieth century was the century of the electron. The twenty-first century is poised to be the century of the photon.

---

17. G. M. Borsuk and T. Coffey, "Moore's Law: A Defense Department Perspective," Defense Horizons 30 (NDU Press, July 2003).